a / o

/

LAURA BYLENOK

a/o

/

LAURA BYLENOK

NEW MICHIGAN PRESS
TUCSON, ARIZONA

NEW MICHIGAN PRESS

DEPT OF ENGLISH, P. O. BOX 210067

UNIVERSITY OF ARIZONA

TUCSON, AZ 85721-0067

<http://newmichiganpress.com/nmp>

Orders and queries to nmp@thediagram.com.

ISBN 978-1-934832-47-9. FIRST PRINTING.

Printed in the United States of America.

Design by Ander Monson.

Cover image: Israel Aguilar Pacheco, "Particle," 2014.

CONTENTS

As I, the I posits the character of the nct-I
—Heidegger

A is A
—The Law of Identity

PREFACE

In her 1985 article on Euclid's lost *Pseudaria*, published in the inaugural issue of *Historia Matematika*, Annette Milosmálmoljac reconstructs the trajectory of that volume as follows. Long-known as a purely didactic compendium of common fallacies, and preserved only insofar as it was casually referenced in works of greater worth or abused in an offhand anecdote, the *Pseudaria* has served as a warning to the amateur geometer against the catalog of tantalizing and banal failures in logic. Any physical manuscript has been until now thought lost to time.

However, Dr. Milosmálmoljac reports the recent uncovering of a partial copy, which she found uncataloged in a basement annex of the Bodleian Library, confirmed and carbon dated to the 5th century AD and consistent with the hand of Proclus. The longer fragment suggests another possible implication and use of the text, collapsing the inferences of truth and false proof and accessing a metaphysical reality beyond the scope and essential understanding of the *Elements*. In particular, the transcription of an equation whose first appearance is otherwise attributed nine centuries later to the *Brahmasphuṭasiddhanta* of Brahmagupta (there expressed in verse), indicates not only an origin but a flexibility and a familiarity with the term: $c/o = o$.

A.

After discovering the equation, I sank into a deep depression. There was no more work that could not be done. It was easy enough to go back to that morning corner, when the snow fell filthy from the slit belly of sky. The telephone hunkered in its booth. The receiver was still warm from someone else's ear, and it smelled sickly, the thick smell of cough drops and cheap musk.

*

Shortly thereafter the anomalies began. It began with the seventh cervical vertebra, the knob at the back of the neck. I was brushing my teeth, my very back teeth, a molar in particular that I couldn't easily reach and I reached the brush back and tipped my head back, eyes on the discolored tooth in the mirror, neck awkwardly bent to see and to brush, to get the bristles in the gap— there it was, the nerve twinge right at the root and my neck popped. I felt my neck. I felt the back of my neck. Something was missing. The skin was warm as it should be, smooth, too smooth, something was missing—no— something was wrong. I let the toothbrush go, let it hang from my cheek, let it trail its string of spit to the sink and I counted the ridges at my neck. One definitely gone.

*

Or it didn't begin. It had always already begun. I was in a neighborhood I didn't know, on my way home and a phone, a payphone on the street began to ring. It may have been ringing for some time when I walked by, I couldn't say. I picked it up. A voice on the line, soft and very clear, said meet me in the hospital waiting room, I have something to show you. I didn't want to say no. It was a frivolous yes, a stranger's yes.

*

When I arrived, the man was pleased. He took my hand and thanked me. He waved me over almost tidily to one of the vinyl chairs and produced from his briefcase a scrap of manuscript. I was so surprised I laughed. I didn't ask his name. He excused himself to make a call and didn't say goodbye. I turned the thing over in my hand—it was the size of my palm—vellum, warm as a pocket, and he was gone.

*

What does it mean to inhabit a body? To know or to believe a body exists—finite, fixed—or for a body to become lost? To belong to an identity, to traverse language, to assume? To say—mine, not mine? To enter time like a numeral, a symbol without a body? Is language enough to sustain a moment of fascination, a moment of real—real—neurons cascading, electric: on off on off. To be on or off. Or what does it mean for a body to inhabit I? To inhabit (—habit) (in—it) language, thought? Once upon a time. Upon. On. On time or in time or out of it? Once upon a time the numeral zero (aught, not, a loop, a knot) was simply a placeholder, and then it was more: it was nothing.

*

I walked by the telephone several times. I made excuses to myself to go through the neighborhood. Each time I passed, I hesitated, nauseated, hoping the phone would ring. When it didn't, I made excuses to myself to call strangers. I picked random numbers and said hello, good morning with all the fervor I could muster. I said I have something to show you. I said apologies, wrong number. I said goodbye, and nothing else.

*

In one version of the story that was all. A thrill. I called a colleague and asked him to take a look. It was a Wednesday morning, when the snow began to fall. He looked up from his papers. When he rubbed his thumb against the vellum I gave him, the lines of his fingerprint lapped their tongues against the zeroes printed there. Let me keep this, he said. I'm not convinced.

*

I went out and I found the phone. It was still morning. I dialed zero, asked for a doctor. I pulled the scrap from my pocket—centuries old, and yet a copy of a copy. I'd kept it with me for weeks now, worrying its edges, pulling it out and turning it over and slipping it back into my jeans. It was criminal I had it, and no one knew I did. The operator connected me with Dr. B. Hyperboles. Must be a pseudonym, perhaps an anagram, though I didn't know what the B. stood for, so an anagram for what I couldn't guess. I decided to walk. The address was unfamiliar, but the receptionist said the neighborhood was close, and there were no taxis on this street.

*

The snow got dirtier as I walked and several times I slipped on the ice, packed by unnumbered dirty boots, dirtier and dirtier, perhaps, I imagined, all on their way to the same overexposed waiting room. It was a mental hospital but he said he'd meet with me. The snow got sharp. I'd like to see it, he said. The snow slick as cartilage. The temperature dropped as noon drew near.

*

In another version of the story I walked down a carless street. I meant to cross. I was careless, I stopped midway, mid-street, on the asphalt faltering, I—the step, half-step, lift of the heel, half-lift without slipping, each movement and its half-life dissolving into the next projected motion, hesitation—slowed. When B. Russell, after centuries of philosophers' and mathematicians' failure to do the same, dismissed Zeno's paradox with a paradox of his own, with a quick diagonal slash through a string of ones and of zeroes like pearls, he was crossing this street. I saw them, his dull pearls, scattered on the snow.

*

In classic science fiction, the consequences of the time machine are predictable. The grandfather paradox, the production of parallel universes, snafus of compossibility, free will held captive in the immutable tower of time, et cetera—each story has its own theory and each theory its own set, or set theory, of infinite and head-scratching consequences. Later, there's a whole generation, unaware of the pun, whining what if they say get out of here kid, you've got no *future*.

*

Noon was my future. Noon drew near like a horse and buggy. Noon came closer, glowering, breathing heavy through its nostrils. Noon shook. Noon shook my hand. Noon tapped its foot on twelve. Noon pulled me in like a fish on a line. O no one O noon I'll dance with you. Noon sang me *a cappella* in common time.

*

This was what went through my mind—what was this what was? There is a point in time, time marked by space, at which was was. Which was—what? In the waiting room for Dr. Hyperboles at noon. There I was. I thought I could get out and I picked up the courtesy phone and dialed zero again. No answer. I continued to wait.

*

I've tried to remember another time—a late fall afternoon, when the blackberries rot the air with ripeness. Afternoon ran, an animal, into the brambles, its skin catching, blood budding and curdling in its fur. I tracked its single path, its berries and burrs I couldn't touch. I lost the scent. I swerved. I tracked an asymptote of cold against its curve.

*

Dr. Hyperboles was excessively calm. He explained the birth defect, overlooked because I had no occasion prior to the incident for radiography. That was his language. Rare congenital absence. I pressed the back of my neck as he continued, as he explained there was nothing he could do, that there was no need for medical intervention. Any pain, he asked. Any stiffness. It just feels different, I explained. We were both explaining, as if we could get something else across. There used to be a knob, and now it's gone, I said. Hm, he said. Any other symptoms and come back in. He wrote me a prescription and walked out of the room.

*

I have made this up. I am made up of what I have made
up.

*

It is the waiting room at noon. The receptionist is out sick, she slipped on the ice, and I'm the only one waiting. Dr. Hyperboles escorts me into a room full of jars of vertebrae. Each jar contains a single vertebra and is labeled in marker with a number. He indicates one and asks me if it looks familiar. The number on the jar is too long and too cramped to read.

*

I took the apartment across the street from the phone. I watched it while I percolated coffee, while I read through the day's mail. I watched the phone while I ate, while I wrote and rewrote the equations. I left my window open and listened for the phone. I gave the number, marked on the box, to friends and asked them to call and listened to it ring. The snow blew in all morning I was listening.

*

The controversy between Gottfried Leibniz and Isaac Newton arose because each claimed to have independently invented the method of calculus. Who published first versus who drafted it first versus who thought of it first versus who had the very first twinge and inkling that something else was going on inside a line. The upshot is L. published first but N. claimed L. stole his idea. I never gave it much thought except as a bit of colorful history to get us all interested early on in what the hell a limit is and why it has anything to do with the rest of our little lives, if it does, as well as to confirm and further train the young and still bright- and not yet glassy-eyed mathematician's sensibility that by writing something down first he can own it and understand it and control it and be credited with genius and glorified in the press and be awarded a signature line in the immortal and gilt-edged registry of history. Whereas if you just cribbed or glossed and in a discerning crib or gloss came across something gut-wrenchingly or exclamatorily new it didn't matter who you were, you were a thief.

*

But there were differences. Oh, always differences. Newton skirted infinitesimals and doted instead on fluents and flux: *Methodus Fluxionum et Serierum Infinitum, i.e.* differential calculus, earned him an interview with Bishop George Berkeley.

BGB: *"And what are these Fluxions?"*

Newton: *"The Velocities of evanescent Increments?"*

BGB: *"And what are these same evanescent Increments?"*

Newton: *"They are neither finite Quantities nor Quantities infinitely small, nor yet nothing."*

*

And then there's the cosmic microwaves, another relic
not yet nothing, invisible and relentless as reason. There
they are, everywhere, still twitching from the blast, the
beginning of time: they are the proof the universe is not
just expanding but expanding and infinite. And if what
the physicists report as pure, as rational as a bubble—
its sud small enough to hold in a palm, in a mind—is
true, it is not a consolation. It doesn't matter what shape
you imagine. A bubble. A berry. Each grows outward,
sickening, into an infinite knot, caught inside my throat.

*

Theoretical cosmologists argue the progression toward infinity is already infinity. Time doesn't have to wait. Time flexes its jaw. Time licks its lips and the bubble, as if it were delicate and particular and mine, as if its little waves—hello, goodbye—were mine, expands. It expands and fills the room, it fills the street, the city, it bloats out like an animal, a body distending on the side of the road, a river of dirty ice, and I am the bubble, the zero, the animal, the kill.

*

A finite system—ours, us—in an infinite field means necessary duplicates. Once all of the possibilities run out, the universe—monoverse, multiverse, verse and refrain—repeats. Must repeat, exactly and inexactly and endlessly. Take, for example, the number of all the particles of the known universe, and call it a. If a is finite, there will be a finite number of arrangements of a. Whereas if space is infinite, there will be an infinite number of iterations of arrangements, and eventually an a exact to the original must repeat. And then again. And again doesn't even begin.

*

There are numbers so large that the full expression of their digits would not fit in the known universe. Skewes's number, Moser's number, Graham's number—recursive formulas reduce them to still-not-quite-manageable notation, but cannot diminish their swagger, their sheer staggering sway. But take the smaller numbers, numbers my body can hold. If I take all the cells of my body and I line them up in a row they will stretch, with a little slack, from the earth to the moon. Then I can change the order, ever so slightly. Then ever so slightly again, one cell at a time. Before I repeat any version, I can change the order a factorial of forty trillion times, a number with its own well-attended cotillion, quadrillions of trailing zeroes and their plus-ones. But all these are dwarfed, reduced from an impractical something to a practical nothing by the face of infinity. It is this practical nothing—the smallest change—that terrifies me most, in the cord I have strung into an order of myself.

*

Every version of the story is inevitable. Every variation. There is no singular vision—every detail will swell, real. There is a version I cross the street. In which I don't hesitate. In which I startle a crow and it flies a diagonal slash across my path. There is a version the crow comes from nowhere, comes out of the snow, in which the particles that are snow are converted to a path and the path is the crow. There is a version the crow strikes the glass of the door of the hospital to which I am walking and the glass fractures its skull. A version in which it is the door that fractures, the beak stuck in the glass, from the impact cracking outward from the wound. There is a version I pull the beak, itself in shards, from the glass. A version in which the blood cools as it soaks into my glove. There is a version in which I hesitate. I don't hesitate. I take my glove off, ecstatic, outside of time, and lick my fingertips.

*

I believe now Leibniz was driven mad by infinitesimals. No one knew this. In the history of calculus—in one of the histories of calculus—the problem—one of the problems—was division by zero.

*

I define the limit of the function $f(a)$ as a approaches zero. I am the limit of the function $f(a)$ as a approaches zero. I am the sum and the limit of the sum I am. The sum of the function of my body I am the hand the pen the my the everything as function $f(a)$ and I am $f(a)$. As a approaches zero—as I am the function approaching zero—zero is approaching increment by evanescent increment. I am an increment. I move my blood, I am the nerve as evanescent as a curve in flux: an outer wall, a zero diving into zero and dividing me alive and null, dull flag ecstatic in the morning light, outside me, wanting in—I want to limn my body from inside—unscrew my blood like a bulb—bare bulb bigger than a white-cold sun—until I am neither nor, yet nothing I touch don't touch me I am.

*

The version I am is unicursal, cursive, a recursive signature, a string to retrace. As if I could return, as if there were a path back out. Crows circle overhead. Brambles full of ones and thorns spill from the earth. The principle is simple. The entrance to the labyrinth is through the mouth.

*

It happened again. I realized there was an extra tooth, maybe a wisdom tooth under the gum, but I had my wisdom teeth out years ago. I could feel it starting to come through. The dentist—Dr. H—placed the x-rays on the light box. Pointed out the teeth, tucked sideways. Said it was a standard procedure to take them out, that I could come back next week. But I had my wisdom teeth out years ago, I said. I had them removed in this office— you took them out, I said. Let's see, he said. Nope. There they are on the x-ray. You must be remembering wrong.

*

Something about the smell of the office—something
like formaldehyde—I was nauseated. It should smell like
fluoride. No, naugahyde. Nitrous. Eugenol and zinc oxide.
Instead it smelled musky, like rotten fruit. I explained to
the dentist, again, patiently, from the beginning. I asked
him if I'd be able to keep the tooth. I wanted to say your
office smells weird, smells like my dream I keep having,
but he was already gone. The receptionist said, see you
Wednesday at noon.

*

Long known as a common mathematical howler, a/o expresses the perfection and the impossibility of my crime. It was a cheat. A loophole like a mobius strip.

*

But this wasn't a loop. There was no backward, no glitchy skip, no return. It was Wednesday at noon. I traced the edges of the scrap of manuscript through the fabric of my pocket. I looked out the window to the telephone on the corner. Clouds clenched like a muscle in the sky, dark as a large intestine. I took out the equation and I turned it over in my palm. I pressed it against my cheek, against my jaw, sore from the surgery. The tooth they extracted was an anomaly—a rootless molar, once decayed, pocked, but unusually luminescent, a misshapen moon. I brought it into the lab, the same lab I had taken the manuscript for testing, and I didn't say where it came from. They said to call today at noon.

*

This is what's going to happen—I'll walk out to the phone to make the call. I'll dial zero for the operator, ask for Dr. B. Hyperboles, head of the lab. I'll ask him what he's found and he'll tell me the tooth is dated to the mid-seventh century, probably from India based on trace residues found in pockets in the enamel, probably buried there until recently and then picked up and cleaned and polished before it found its way into my hands. Probably.

*

I won't find a taxi, so I'll walk and as I'm walking the temperature will suddenly drop. I'll slip. The ice will be sharp and I'll twist my neck. When I'm waiting in the examination room, I'll rub the back of my neck, certain there's something missing. Something will be missing.

*

To the average arithmetician, of course, a/o is undefined, an educated way of saying the expression has no meaning. Berkeley, in his scathing treatise against Newton's calculus, called a's over zeroes the ghosts of departed quantities, which might be the only math he's remembered for, a clever and an elegiac turn.

*

I believe now Berkeley was the secret founder of a little-known sect called the Protectorate of the Zero. After the discovery, they came by my house with hole-punched cards and no names and political surveys and pamphlets about MENSA. I thought it was a joke.

*

Can quantities have ghosts? And who would they haunt?
I check my watch.

*

One morning when I passed, the phone was, briefly, animal. A deer, perhaps. A corpse. I stopped and picked up its jaw and placed it against my mouth. Intestines curled away from the handset in a single cord. Behind the static I heard, I listened hard, a message. I couldn't make it out. A woman was distressed, she asked me to meet with her. She mentioned the hospital. I pressed my finger on the hook to clear the line.

*

In the theory of quantum entanglement, apparent paradoxes—anomalies—occur when the behavior of a given set of particles cannot be disentangled. Of course this violates just about every one of Newton's laws: a particle spins clockwise and counterclockwise, an electron has a positive charge, a body is both living and dead. It is a matter of measurement and perspective that contradictory elements appear juxtaposed, collapsed.

*

Some of my body felt young and some of it felt old, and I knew none of my body was mine. On Wednesday I slipped and broke my wrist. The doctor scratched his beard and said it was an anomaly, the bone was deteriorated, its mineral density that of an eighty-five year old woman. The rest of my bones appeared normal.

*

I woke up with Leibniz's liver next to mine. This wasn't a dream. I slept in and woke up right at noon and knew. I could feel his cells, his zeroes, creeping next to mine. He began to filter my blood. I called and asked Dr. Hyperboles for an appointment later in the day but he was booked.

*

When I walked in, Dr. Hyperboles stood in the room with a scalpel in his hand. The first cut in an autopsy is a Y incision to the chest, and he stood above the Y in his chest like a mouth unable to find the form of a word. He gaped. His chest gaped. He gripped the scalpel hard and didn't look up when I stepped through the door. There were rows of stainless steel tables in front of him, more than I could count, each with a version of himself in a different state of dismantlement. Many looked identical, but I could tell them apart. Inside each, a single particle had begun to grow astray—a space, an a's indefinite article I knew I could pronounce.

*

Any number divided by itself is one: 10/10 = 1, -173/-173 = 1, any *a* over *a* = 1, except for zero. 0/0 = 0 gets off on a technicality—the jury's unable to convict. To say a/0 is to say 0/0 is to say any number and all numbers is to say the real trick about a/0 is to realize zero is not a number at all, but time, and if time is infinitely divisible, it must also be infinitely expandable. But to describe 0/0 is impossible without the equation, and I will never tell you what I've found. I can never want to tell you. I can only say I've entered zero—that zero has taken me over and I am not one.

*

It was then I realized I couldn't change the frame. I wasn't bothered by the winter noon until I knew I couldn't change it. Then, I couldn't help but want to change it, to dream about late summer and late afternoon. It was noon. A winter noon, a body bag. I can't remember what it was like, the smell of blackberries ripening in the sun.

*

Strangers still stopped to use the phone. A woman walked up to it and dialed. I watched her cradle the handset against her neck, and with it there she—I didn't want to watch. She took off her gloves, seemingly, carelessly. There was something peculiar about the way she looked up to the window. She looked away and when I looked again I saw she was missing a hand. When she walked away she left her gloves there, fallen in the snow.

*

Even as it stayed the same the flux got worse. I could feel a blackberry in my brain, hardening to a tumor. Perhaps it started with a single cell, a single cell that couldn't die and in turn programmed cells that couldn't die, and the tumor began. It grew noon by noon and I could feel it grow. I watched the phone from the window, silent, waiting for the snow to come. This had to stop.

*

The vellum grew backward, returned to skin. The diagonal slash in the stomach of the animal spilled coils of cord. I held the knife at an angle and pried the organs up. I pried my fingers into the soft tissue and felt its pulp under my nails. It was warm. I was so surprised I laughed. I didn't ask where it came from, I just pulled it out, the manuscript, and turned it over in my palm.

*

As if I could return, fall backward. The principle is simple. To go back, back falls always forward, and I have been falling, fall swelling forth until it is all, is null, a cipher of numerals, of nerves, until I am the fall—the snowfall in the street. The footfall in a slash of dirty snow.

*

It was Wednesday at noon and from the rooftop of the hospital, the city reflected back the heat and smog over the buildings to contort the air so I couldn't make out the mountains to the south, and as I looked across I couldn't believe it but the first wet flakes unfolded from the air and sort of fluttered down; and I stood up as, all around, it was as if someone had torn open the sky and through the smog suddenly the white mountains seemed close enough to throw a pebble at (I imagined a small *ping* against the great glacial ice), and the shingles and the rooftops and the mountains seemed part of a backdrop painted with falling brushstrokes; and over the patio wall, I looked down into the street, curiously empty except for a spot of red, and the snow was now dashing this way and that above an object (yes—a sweater someone dropped on the sidewalk below), obstructed by the snow with its unchoreographed movement, its cold ecstasy of triangles, spirals, its apparition of empty cones; and I thought, but it won't stick; and I opened my hand to catch the flakes and they melted the moment they touched my skin; but there, below, the red sweater caught the snow, and the snow mottled it white; and but how, I thought, could it not melt; and there, in the seven stories down, with the snow skating downwards in a slow volley, I was lost in a movement outside of myself but which I still felt—in the way my arm moved up, unconsciously, inevitably, to pull my own blouse tight around my neck, as if I were cold; and the sound of a police siren wound through the streets; and into that wail the snow receded; there was no sweater, no snow; a woman—a patient, who might just as easily have slipped and not jumped—had fallen from her window; there was her body; and sirens; and heat coming off the pavement below.

*

And there I am in the late September light at half past seven and it's still golden in the thorns when I shoulder through to reach each falling fermenting handful of blackberries, with each handful falling apart, lobes black bright as a prick of blood, swollen as a tick feasting, bursting as I pluck, suck my knuckle, then taste one after the next after the next, warm as an earlobe in the last sun—then look up—I've lost myself in bars of light and sky now dusted with stars like yeast, alive and leavening the air until it rises and the afternoon is gone.

ACKNOWLEDGMENTS

Thanks to *Valparaiso Poetry Review*, where an excerpt in an earlier version first appeared. For essential guidance and support, I am grateful to my teachers, especially to Jackie Osherow, Kate Coles, and Melanie Rae Thon for unwavering honesty and generosity. Many thanks to Claire Wahmanholm, Susannah Nevison, and Sara Eliza Johnson for sustaining friendships, and to Ed Trefts, Rachel Levy, Tessa Fontaine, Adam Giannelli, and Adam Weinstein for first reads and sharp insights. Sincere thanks to New Michigan Press and to Ander Monson for supporting this work. And for enduring encouragement, I am grateful to my family, and to Israel for his faith in the infinity of little things.

LAURA BYLENOK is from Seattle, Washington, and she holds degrees from the University of Washington and the Writing Seminars at Johns Hopkins University. Her poetry can be found in *Pleiades*, *North American Review*, *Guernica*, and *West Branch*, among other journals. She is a Vice Presidential Fellow at the University of Utah, where she is pursuing a PhD in creative writing. She lives in Salt Lake City.

/

COLOPHON

Text is set in a digital version of Jenson, designed by Robert Slimbach in 1996, and based on the work of punchcutter, printer, and publisher Nicolas Jenson. The titles are in Futura.

/

NEW MICHIGAN PRESS, based in Tucson, Arizona, prints poetry and prose chapbooks, especially work that transcends traditional genre. Together with DIAGRAM, NMP sponsors a yearly chapbook competition.

DIAGRAM, a journal of text, art, and schematic, is published bimonthly at THEDIAGRAM.COM. Periodic print anthologies are available from the New Michigan Press at NEWMICHIGANPRESS.COM/NMP.